# The Meadows Of Heaven

### Austin P. Torney

# Copyright

© 2012 Austin P. Torney

# The Meadows of Heaven

We, of the highest consciousness ever known
And of the most versatile form that's been shown,
Reside as consequent beings in this Earthly realm,
Possibly the most fortuitous creatures
That the universe has ever wrought.

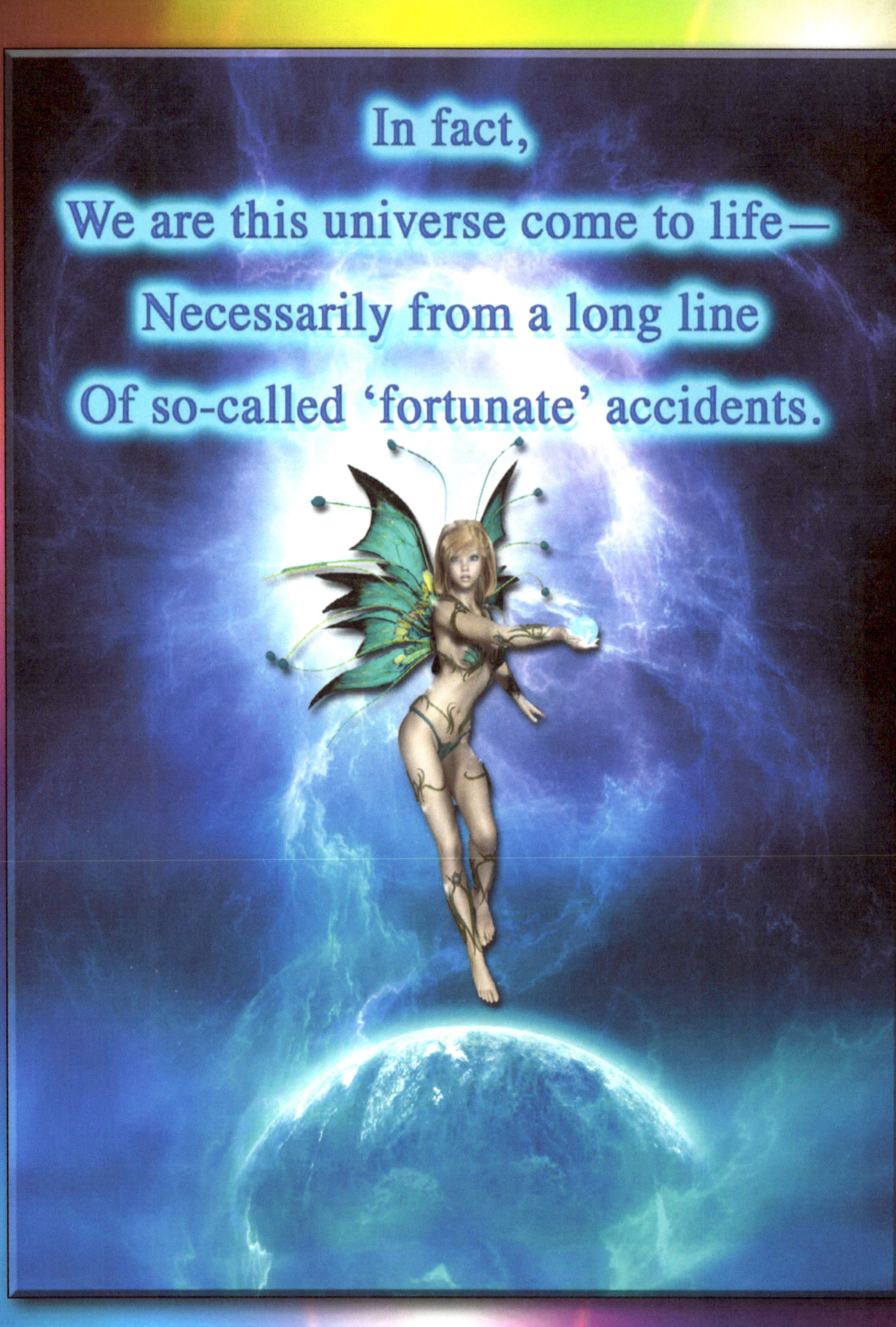

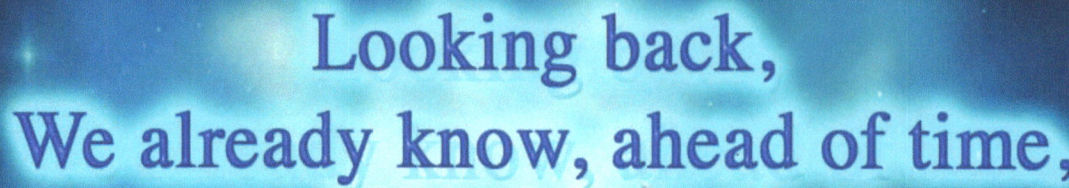

Looking back,
We already know, ahead of time,

That we will 'discover'
The many 'happenings'
That made us possible.

All this we know and expect,
Because we are here.

Perhaps,
in some other 'wheres',

Junkyard universes
litter the omniscape,

For they flunked,
failed, and miscarried—

A quadrillion trillion
universes broken down

For every one that worked
to any extent at all.

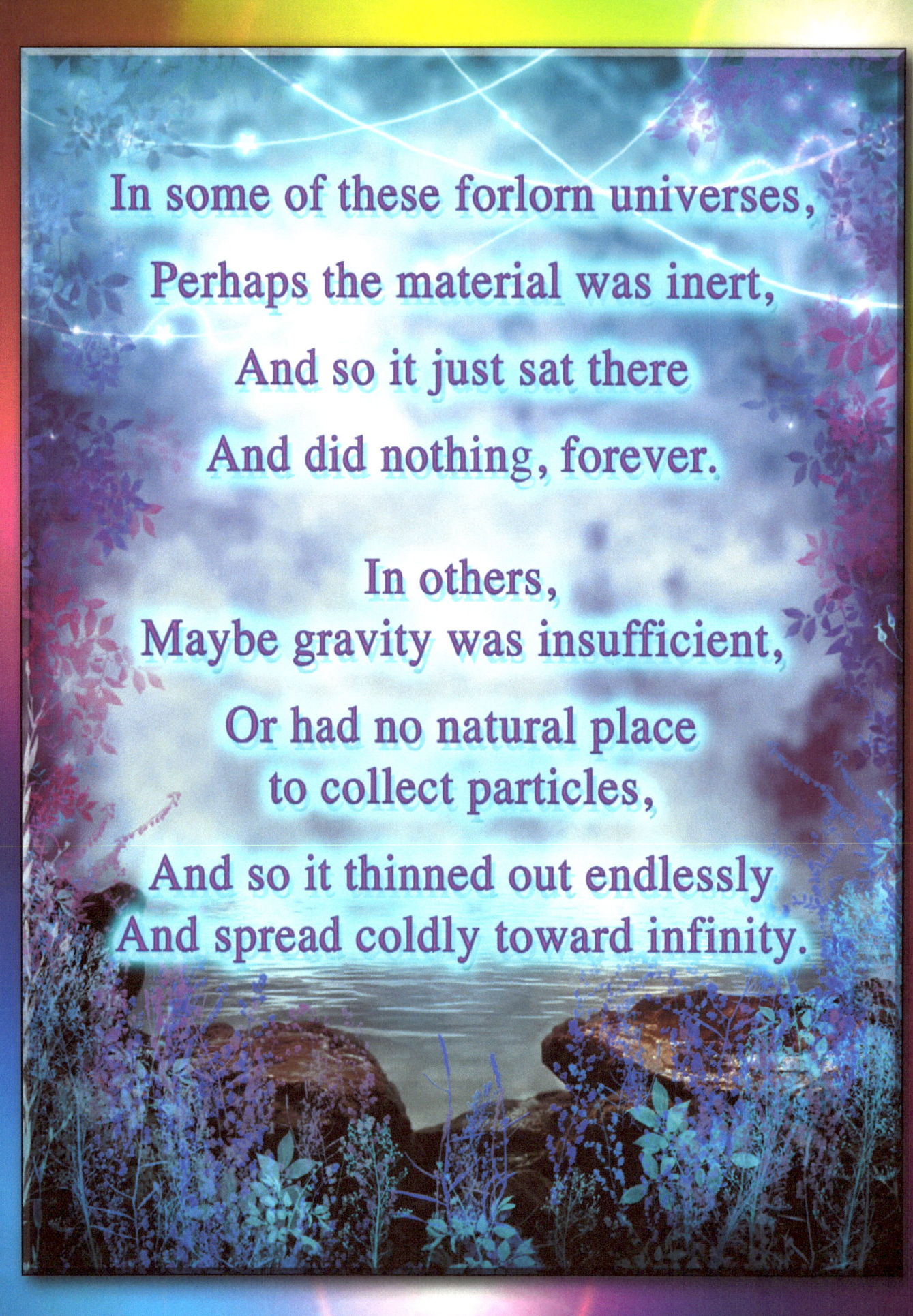

In some of these forlorn universes,

Perhaps the material was inert,

And so it just sat there

And did nothing, forever.

In others,
Maybe gravity was insufficient,

Or had no natural place
to collect particles,

And so it thinned out endlessly
And spread coldly toward infinity.

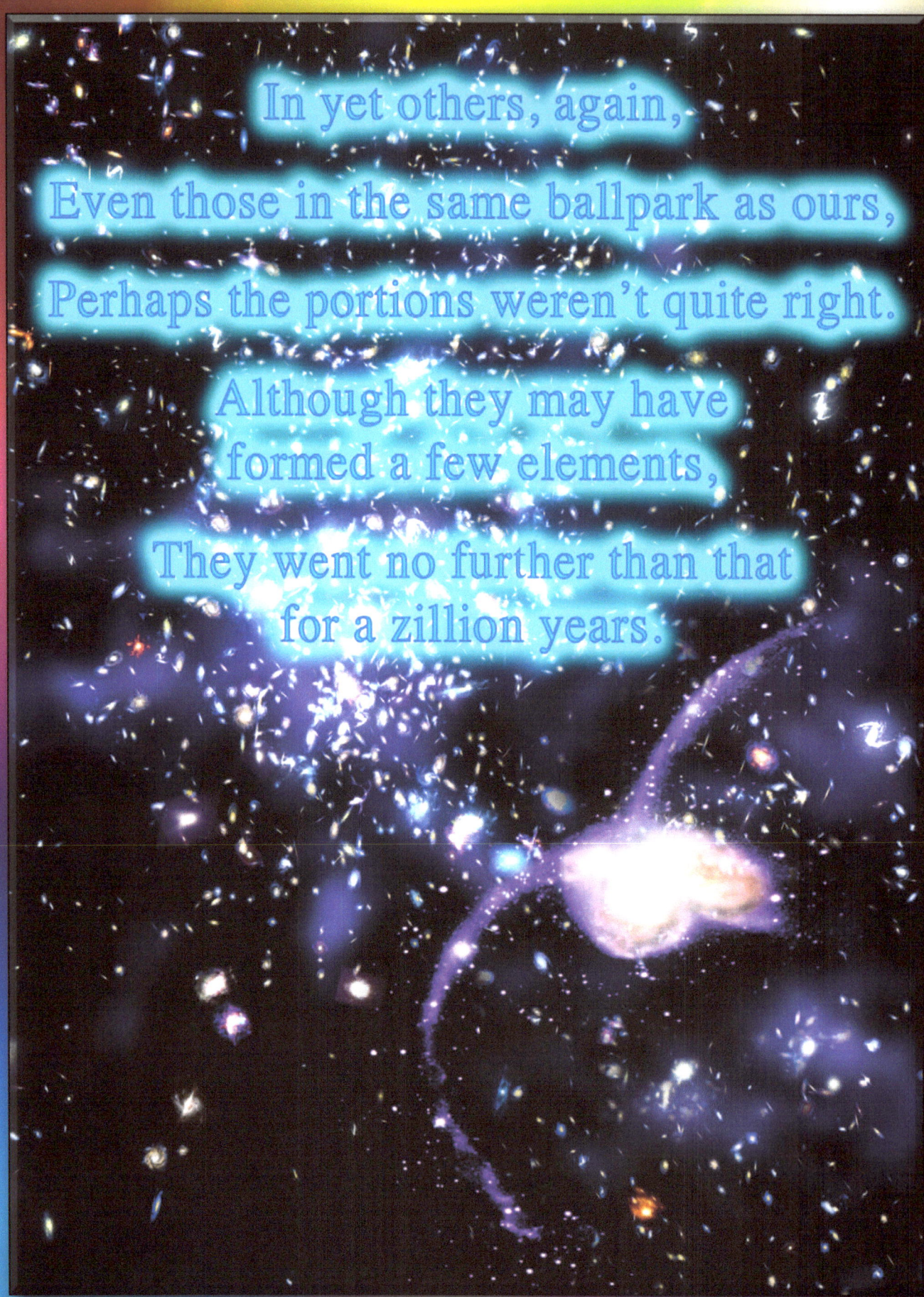

In yet others, again,
Even those in the same ballpark as ours,
Perhaps the portions weren't quite right.
Although they may have
formed a few elements,
They went no further than that
for a zillion years.

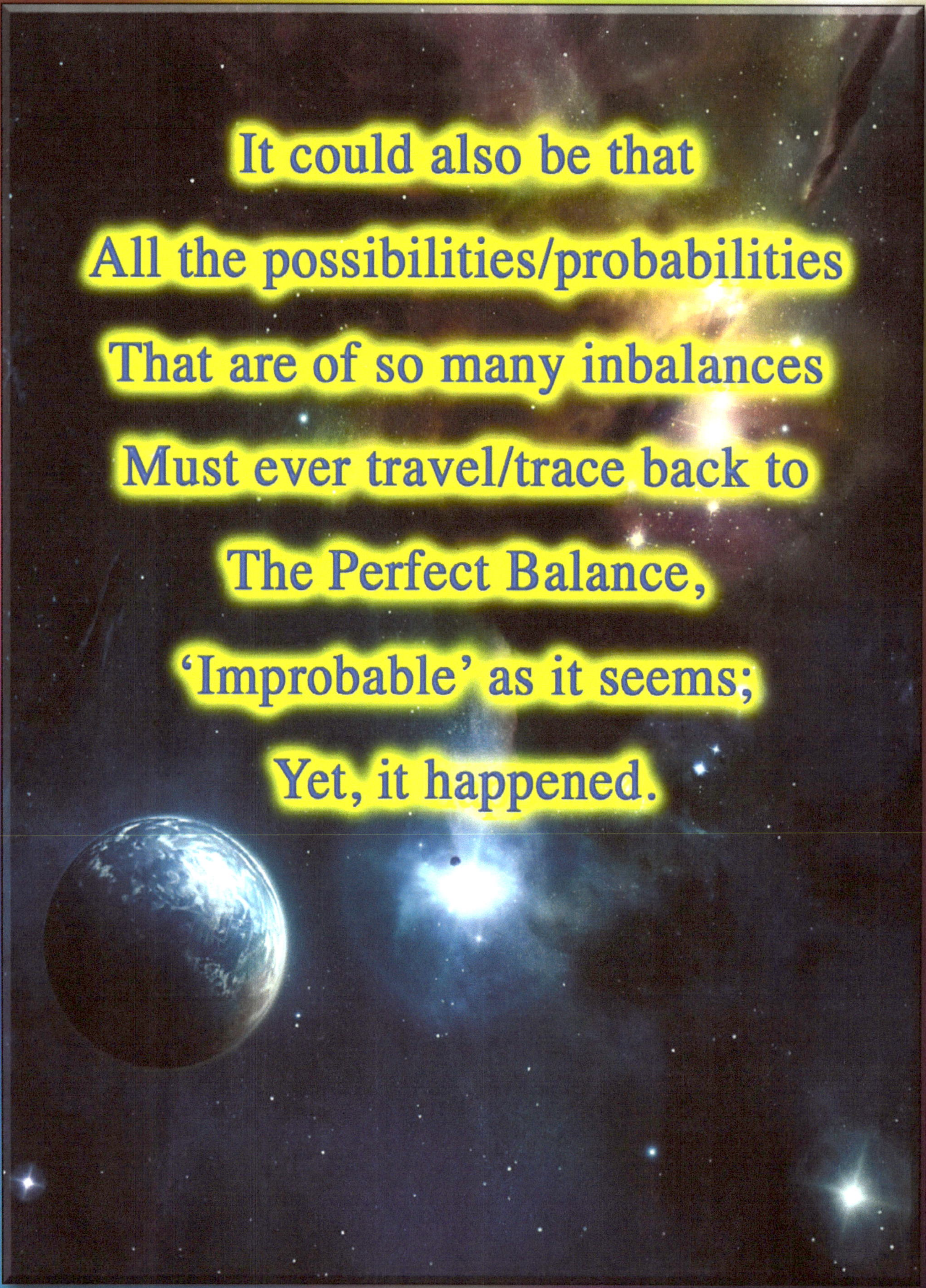

It could also be that

All the possibilities/probabilities

That are of so many inbalances

Must ever travel/trace back to

The Perfect Balance,

'Improbable' as it seems;

Yet, it happened.

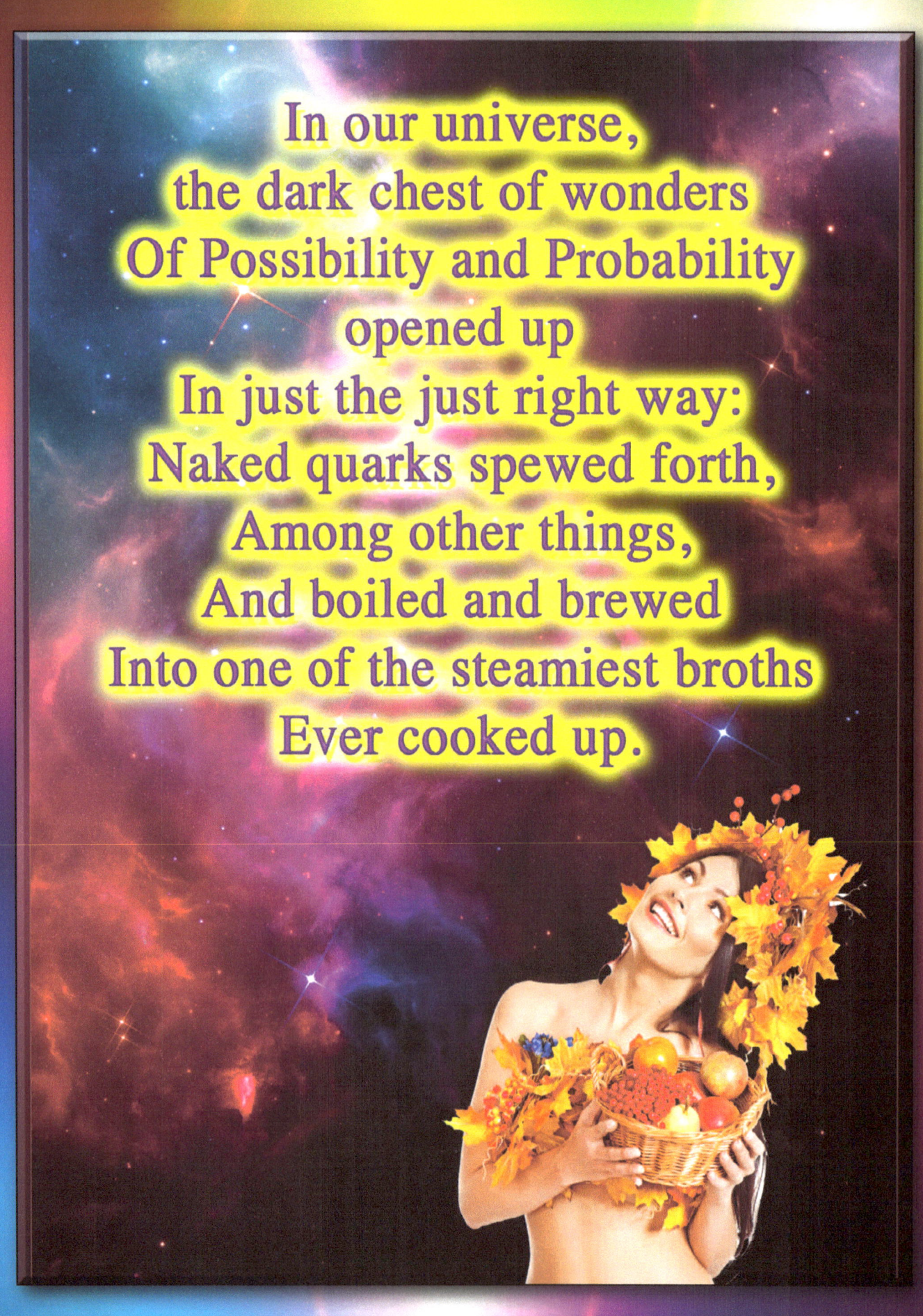

In our universe,
the dark chest of wonders
Of Possibility and Probability
opened up
In just the just right way:
Naked quarks spewed forth,
Among other things,
And boiled and brewed
Into one of the steamiest broths
Ever cooked up.

They somehow simmered and combined

Into the ordinary matter
Of protons and neutrons.

Quite independently,
By some unknown means,
Dark matter/energy arose, as well,
In just the right mix,
and, luckily, too,
Some very long filaments,
Called cosmic strings,
Formed and survived long enough
To be useful as collection agents;
They were merely imperfections,
As in an unevenly freezing pond—
A kind of a cooling flaw.

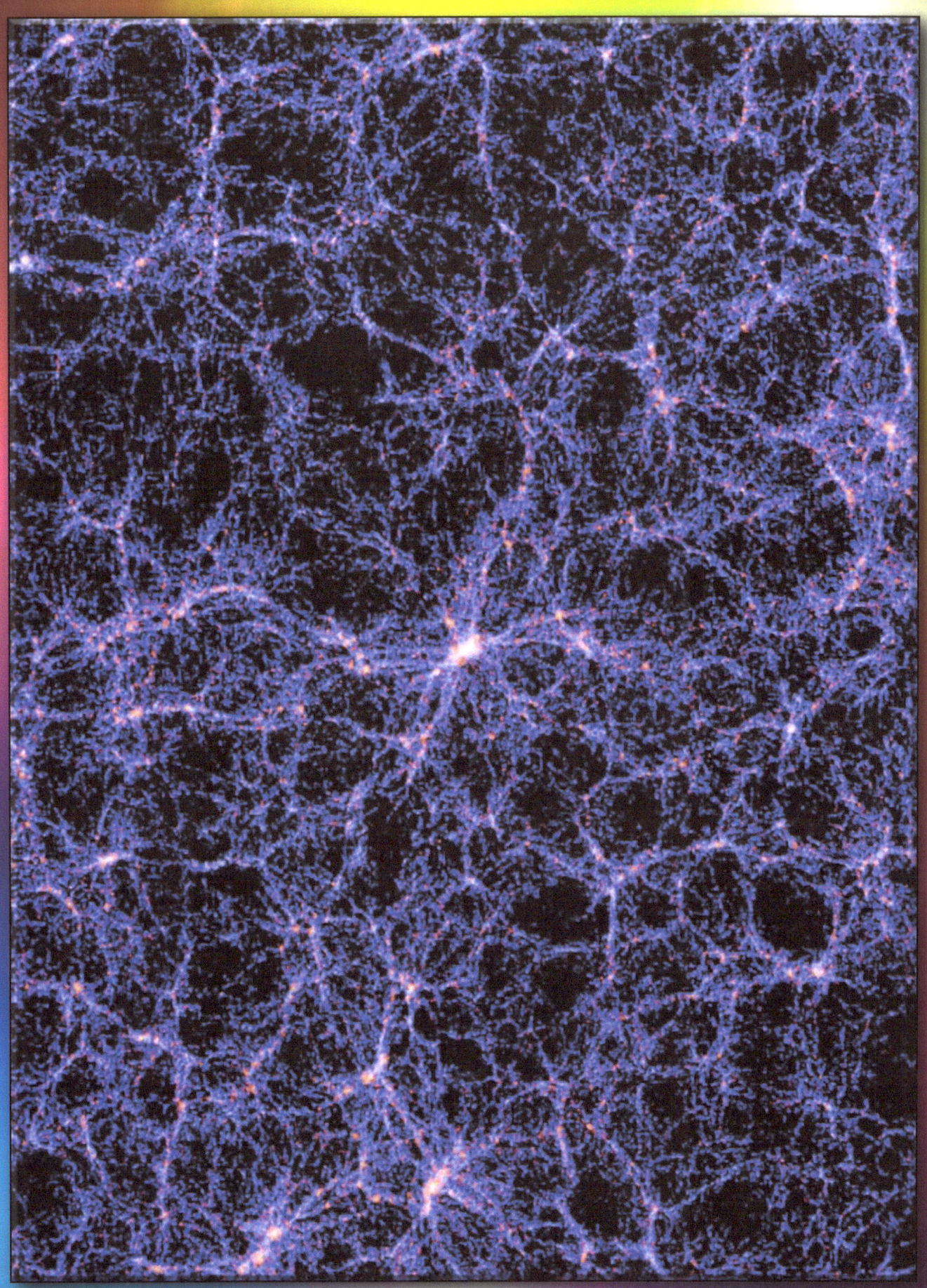

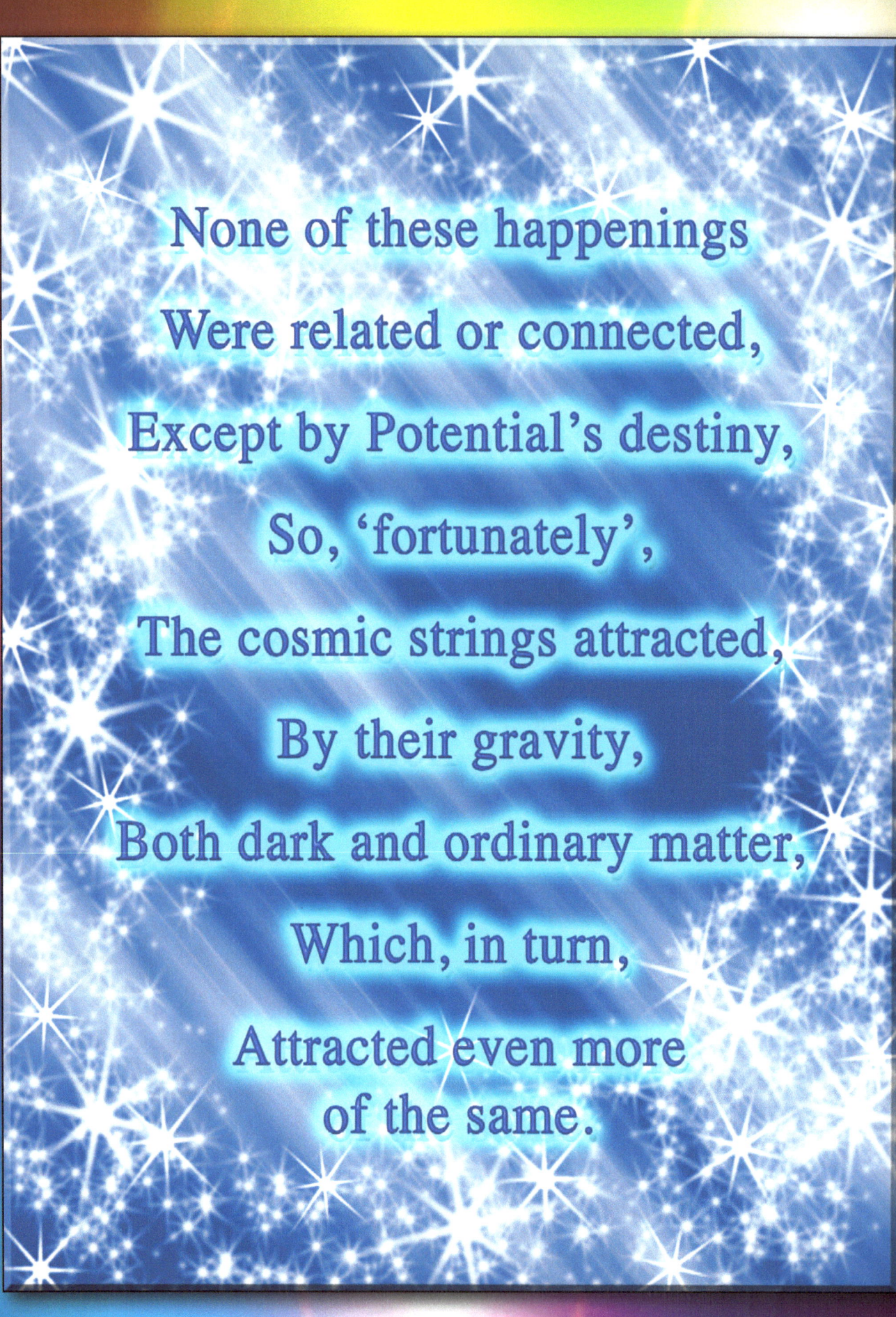

None of these happenings

Were related or connected,

Except by Potential's destiny,

So, 'fortunately',

The cosmic strings attracted,

By their gravity,

Both dark and ordinary matter,

Which, in turn,

Attracted even more
of the same.

These pearls of embryonic galaxies arose

And were strung along these cosmic necklaces,

As can still be noted today.

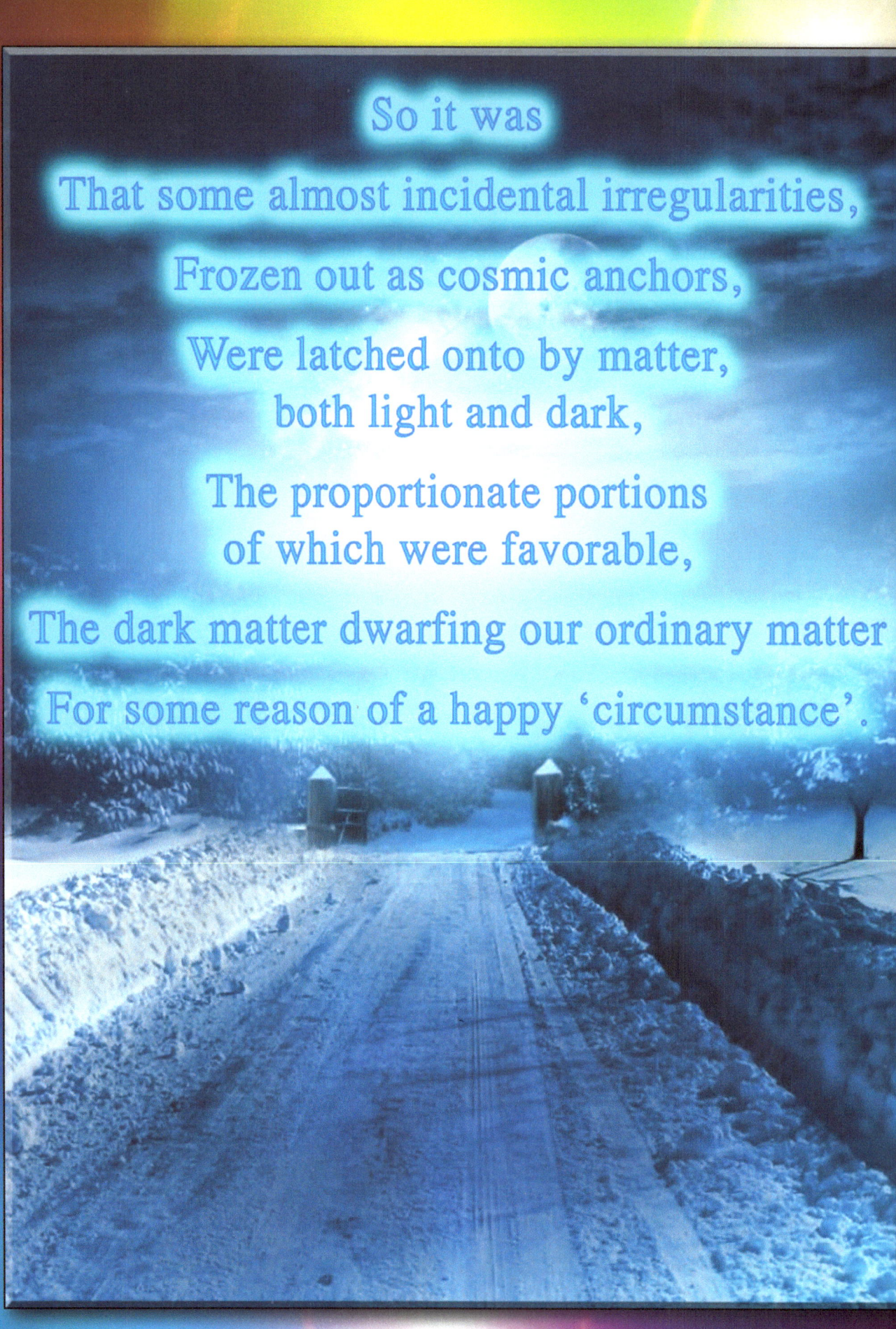

So it was

That some almost incidental irregularities,

Frozen out as cosmic anchors,

Were latched onto by matter,
both light and dark,

The proportionate portions
of which were favorable,

The dark matter dwarfing our ordinary matter

For some reason of a happy 'circumstance'.

'Fortuitously', as well,

Anti-matter, if there ever was any,

Did not fully cancel out the uncle-matter.

The universe could not foresee any of this

In and of itself's fundamental substance(s),

For, if it could have,

Then we'd only have the larger problem

Of how the foreseer

Could have been foreseen,

Ad infinitum…

So, it could have been
like the 'trying out'

Of all possibilities in superposition…

A brute force happening

Of every path gone down,

Or at least everything

Having to happen sometime.

We know much of the rest of the story
Of how the stars and their supernovae
Created the light and heavy elements
Which combined into molecules,
Which, 'auspiciously',
Became able to replicate themselves, as DNA,
And progress to make cells, tissues, and life.

And then there was the luck of oxygen,

A mere waste product of photosynthesis

By bacteria, and later, plants,

That could fill the lungs,

As well as build an ozone layer of protection

From the harmful rays of outer space.

It bears all the hallmarks
Of 'everything' at work,

Although quite probable
If the Potential/Possibility
Had it all 'worked out'.

Dinosaurs roamed the Earth
For over two hundred million years—
Imagine the length of that time.

They were supreme and invincible—
The kings of all the Earth 'forever',
On land, sea, and even in the air—
Heading towards forevermore and beyond,
But…

Dame Fortune once again intervened
When the asteroids or volcanos,
Or some such catastrophe,
Finished off the dinosaurs,
As well as 90% of the existing species.

This event left a vacuum
In which new species could thrive.

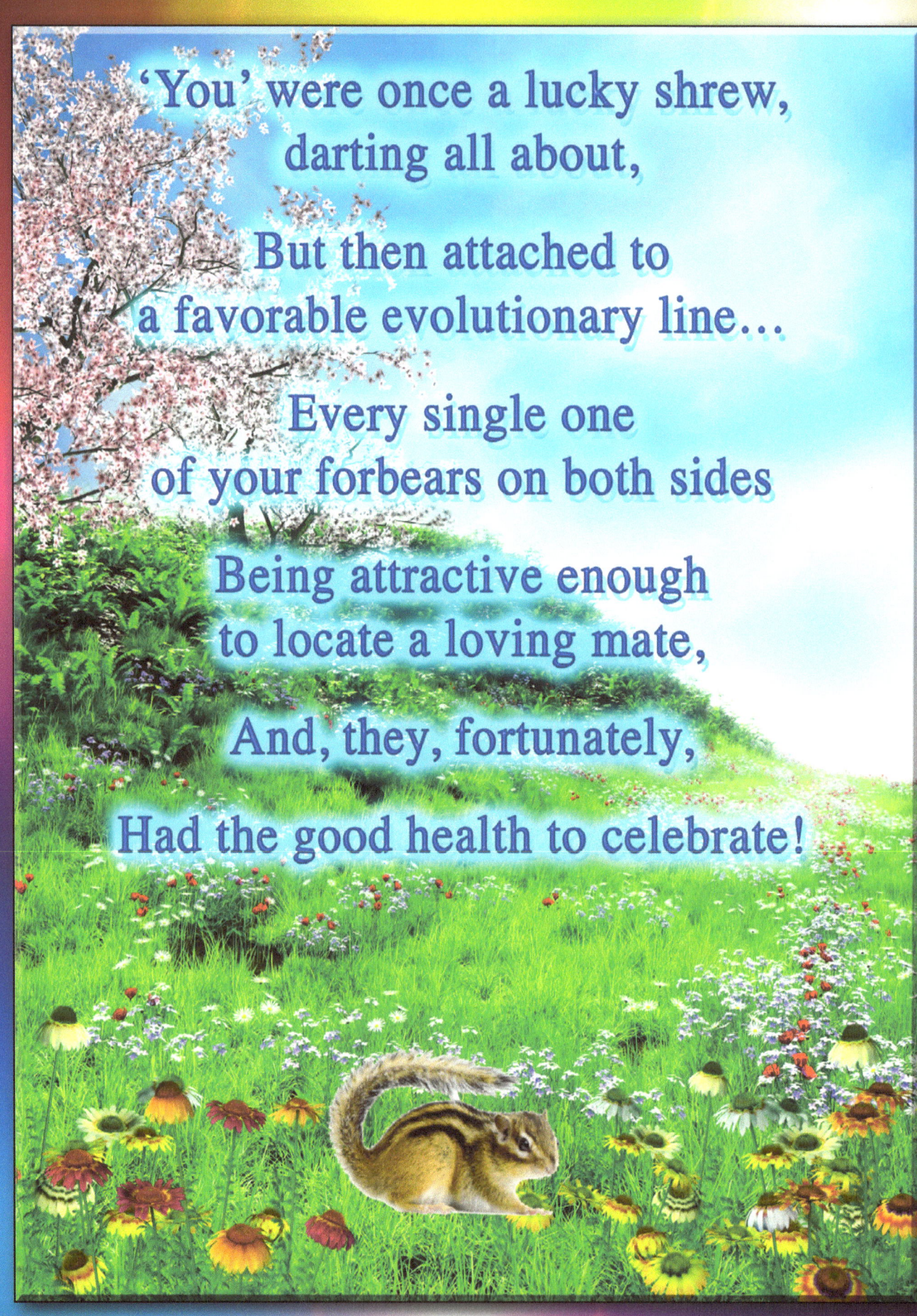

'You' were once a lucky shrew, darting all about,

But then attached to a favorable evolutionary line…

Every single one of your forbears on both sides

Being attractive enough to locate a loving mate,

And, they, fortunately,

Had the good health to celebrate!

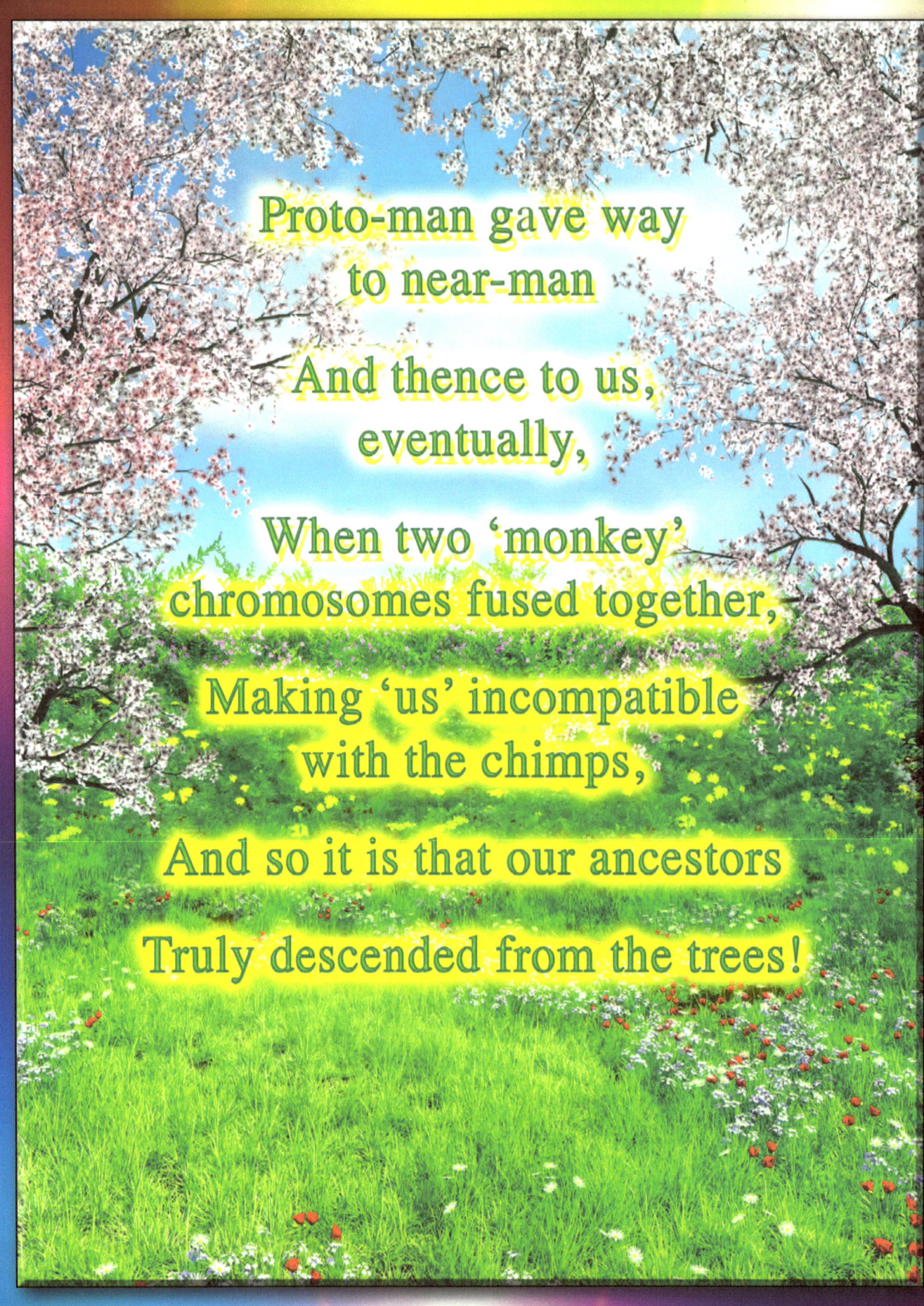

Proto-man gave way
to near-man

And thence to us,
eventually,

When two 'monkey'
chromosomes fused together,

Making 'us' incompatible
with the chimps,

And so it is that our ancestors

Truly descended from the trees!

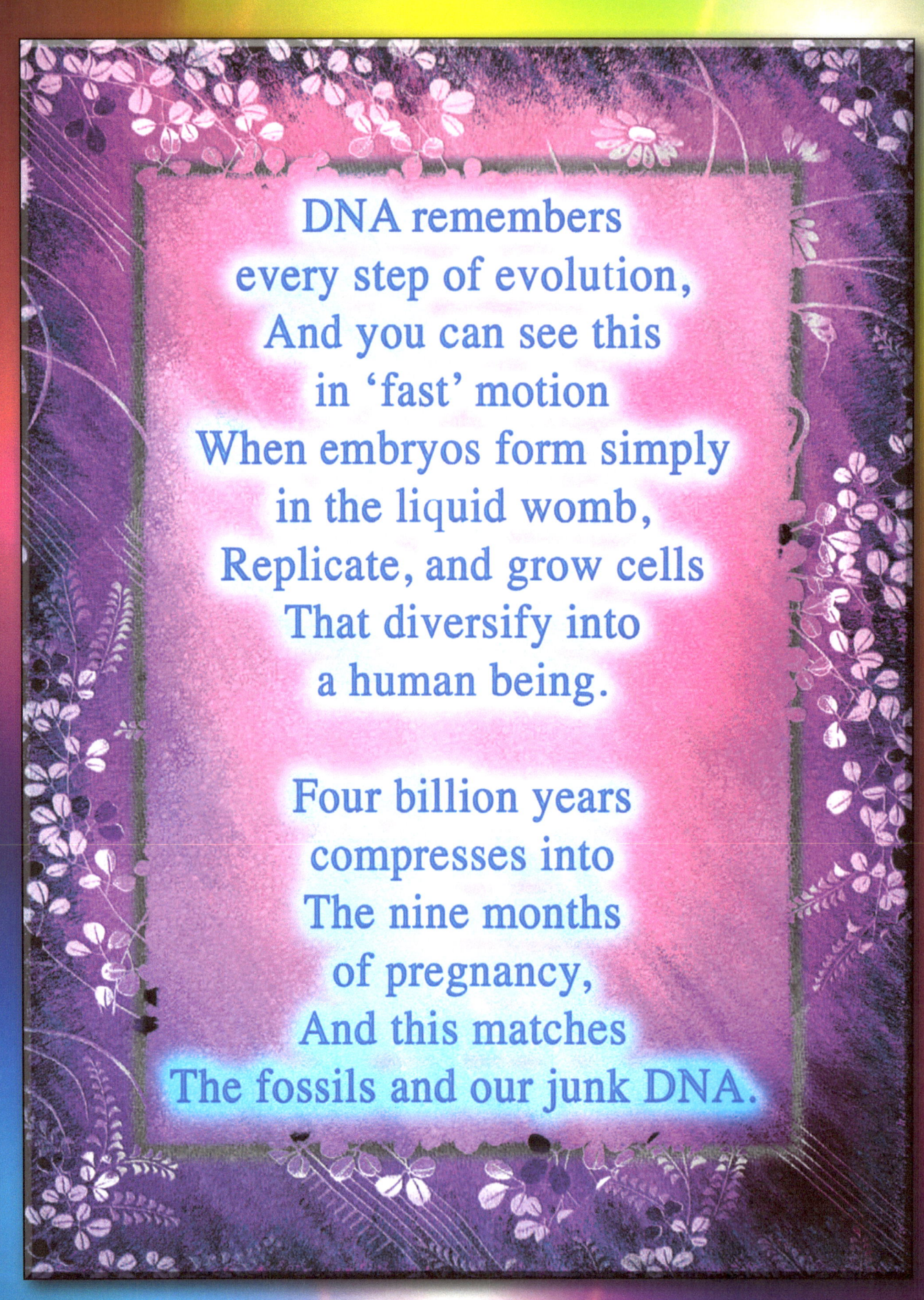

DNA remembers
every step of evolution,
And you can see this
in 'fast' motion
When embryos form simply
in the liquid womb,
Replicate, and grow cells
That diversify into
a human being.

Four billion years
compresses into
The nine months
of pregnancy,
And this matches
The fossils and our junk DNA.

Our higher consciousness
Was the crowning glory;
We had won the human race,
The be all and end all; the grand prize
Of the universal lottery.

So, then, hail, and good fortune,

Fine fellows and ladies,

And welcome all of you

To the Meadows of Heaven—

The highest point of all being,

Although we are surely still in our infancy.

The path chosen
by Potential ends here,
With our consciousness.

The further design
And the role of mankind
Is now in our hands.

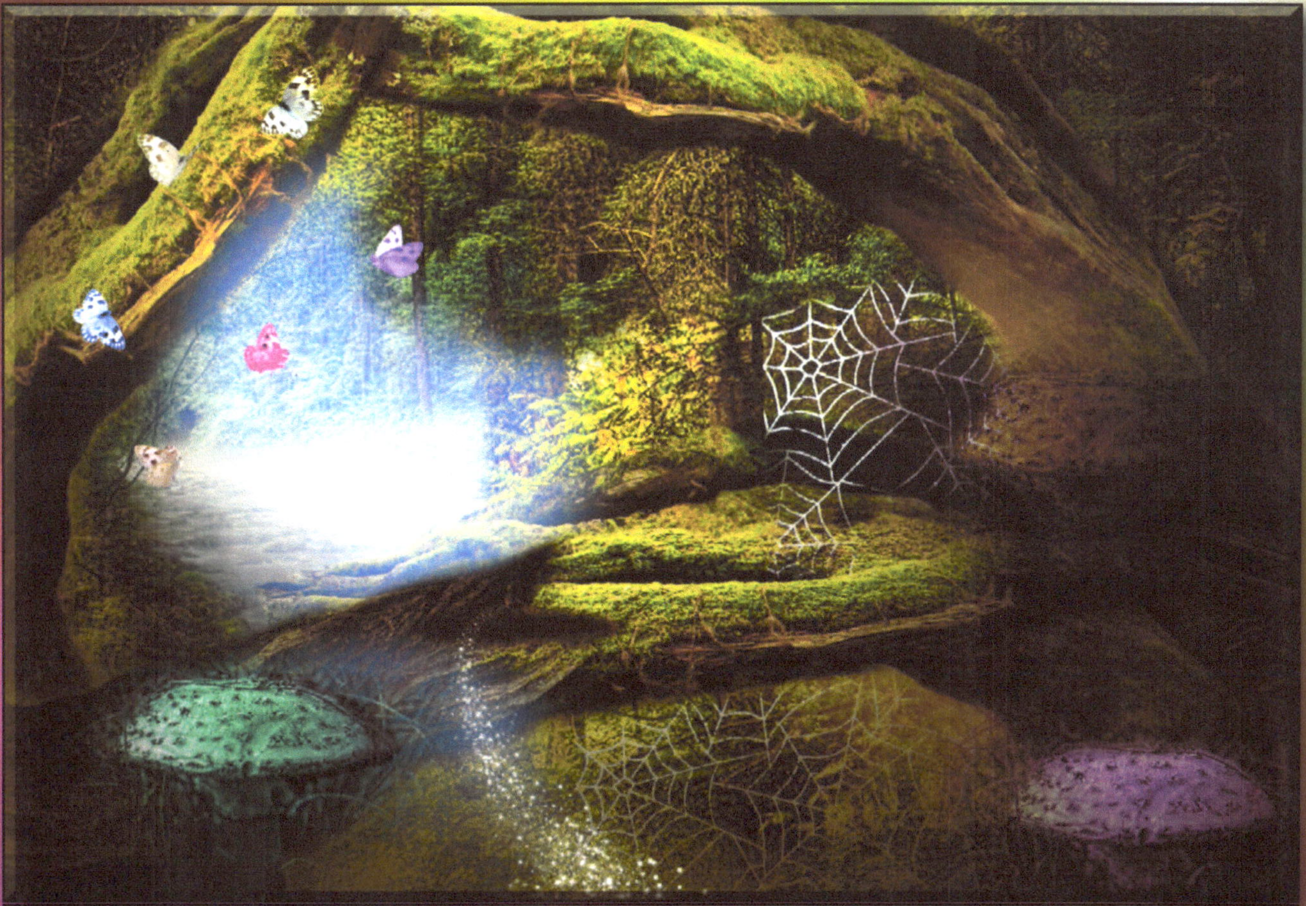

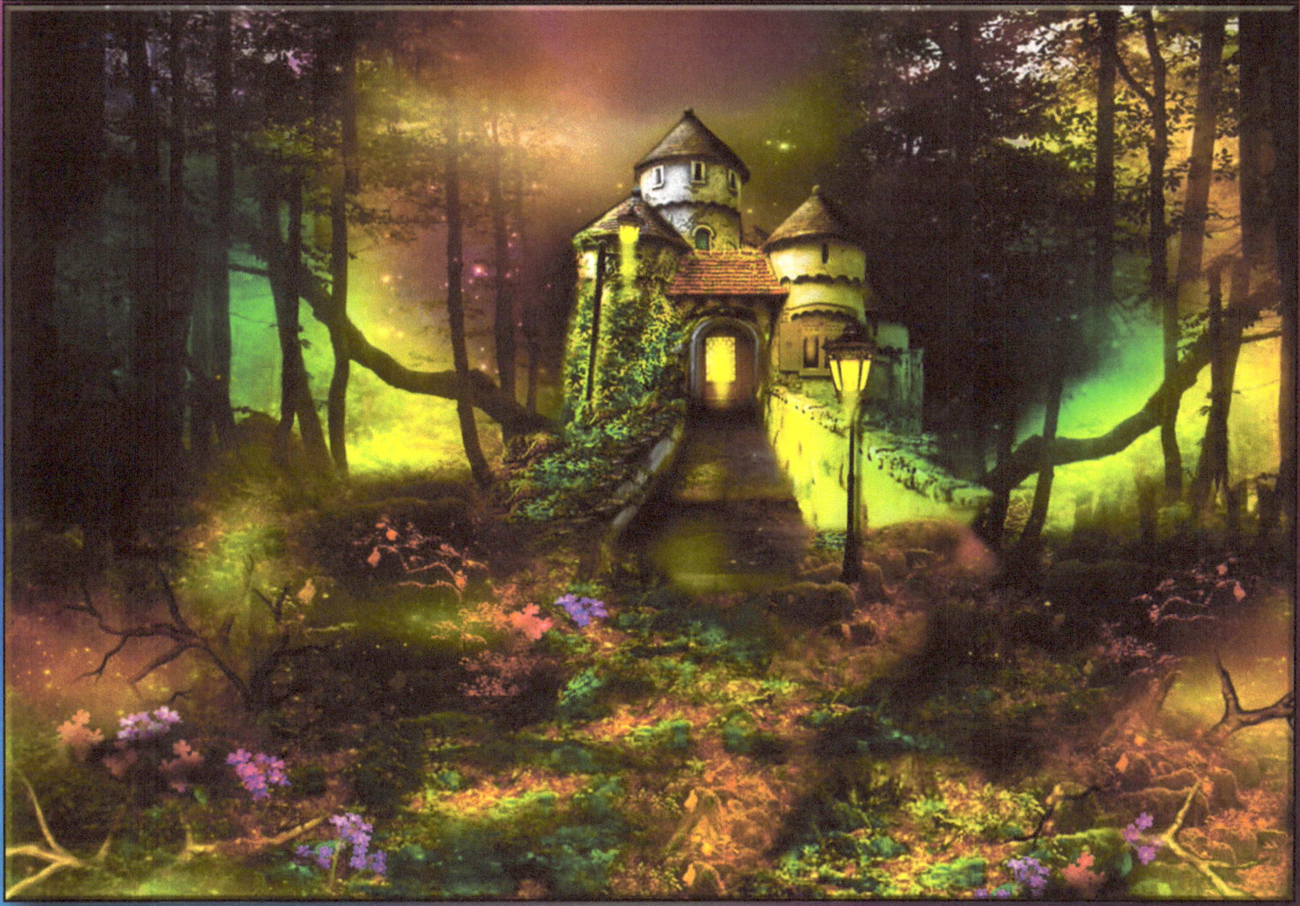

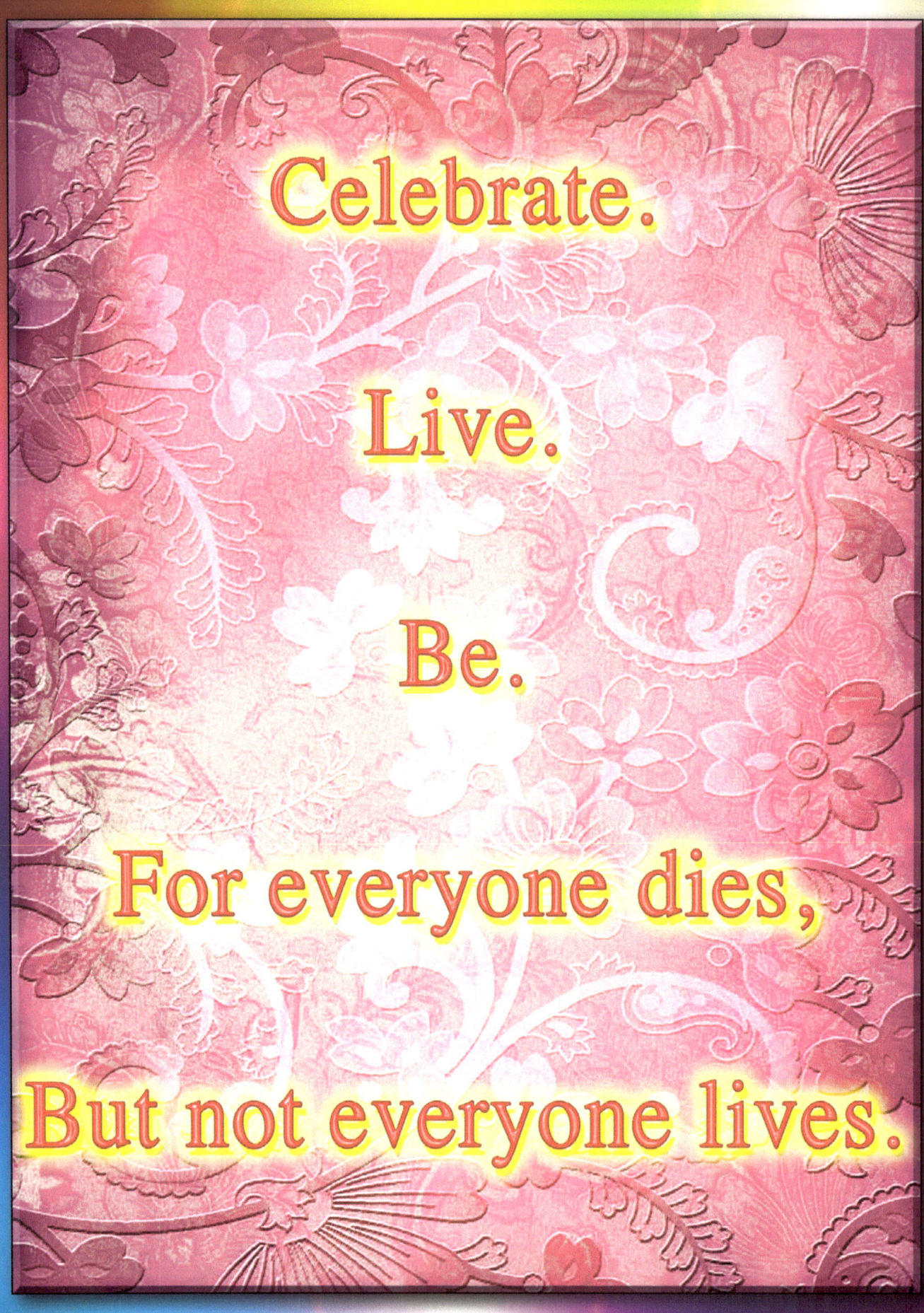

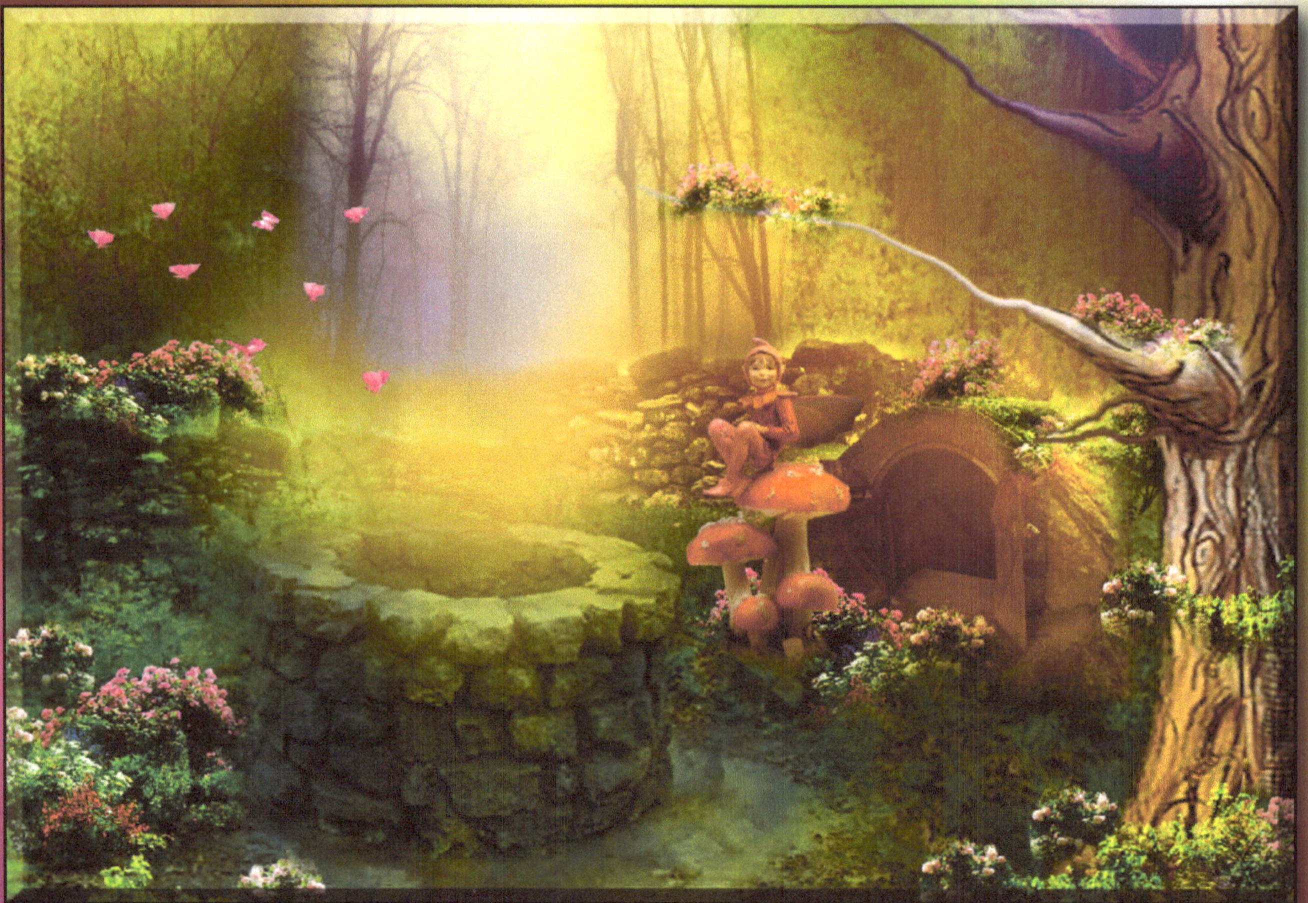

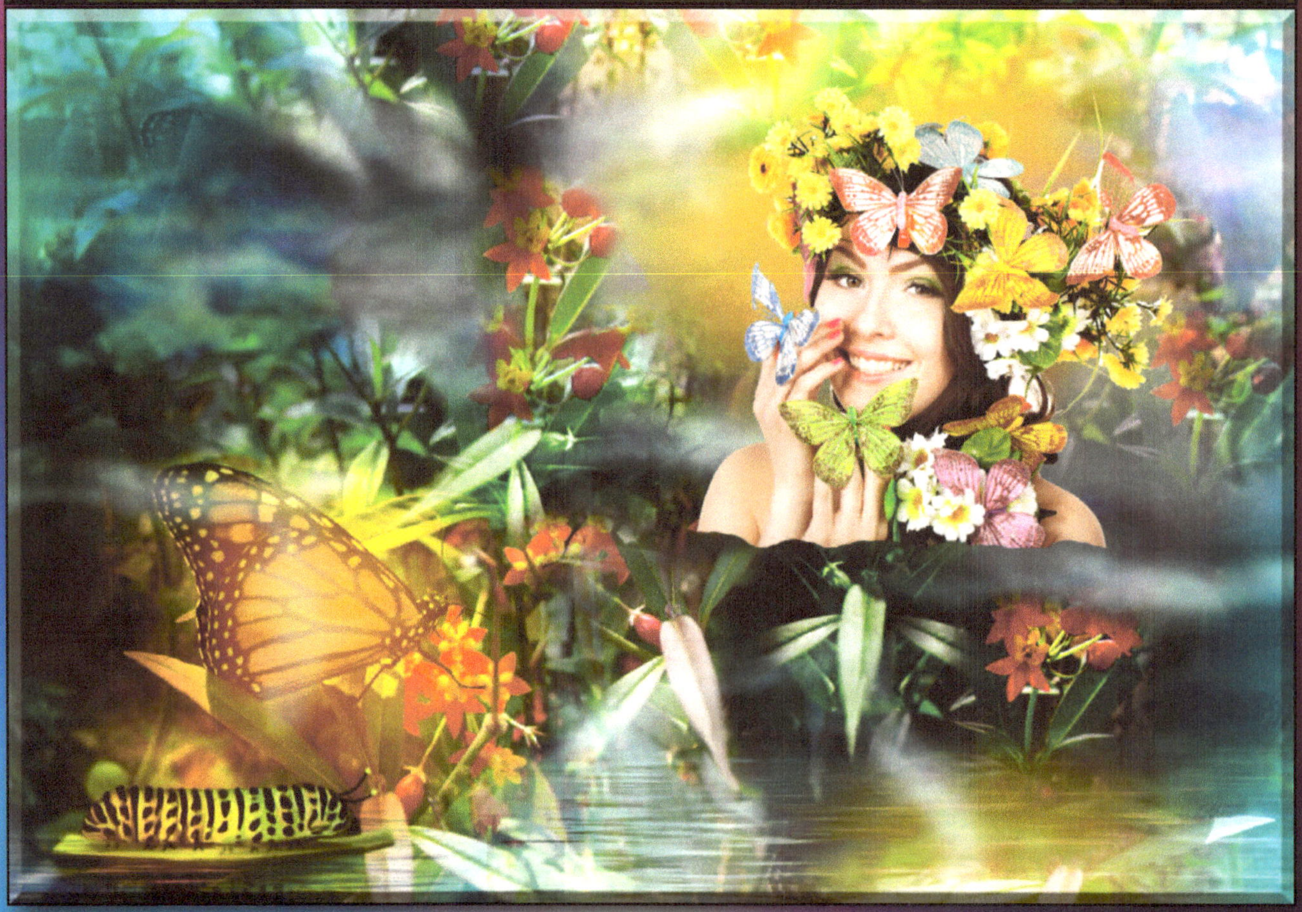

www.ingramcontent.com/pod-product-compliance
Lightning Source LLC
Chambersburg PA
CBHW050735180526
45159CB00003B/1233